To dear Gill + Sandy —

An appreciation for the wonderful joy you gave by your visit to our home, Herousgate, Waikanae, February 2006

Mac + Juliette Morgan

Tales of Waikanae Estuary

Tales of Waikanae Estuary

KAPITI COAST WELLINGTON

MICHAEL PERYER

Steele Roberts
Aotearoa New Zealand

Thanks to the Creative Communities Scheme of Creative New Zealand
for a grant to assist with this book.
Photo on page 1: Waimanu lagoon, by Eileen Thomas
Designed by Central Media Ltd, Wellington
Printed by Thomson Press, India

© Text: Mik Peryer 2004; Photos © the contributors

Published by Kapiti Bird Tours in association with:
STEELE ROBERTS LTD
Box 9321 Wellington Aotearoa New Zealand
phone 04 499 0044 • fax 04 499 0056
info@steeleroberts.co.nz • www.steeleroberts.co.nz

ISBN 1-877338-22-2

Contents

Introduction by the Mayor	6
Waikanae Estuary	9
Birds in our backyard	10
The little community	16
Royal spoonbill	19
Variable oystercatcher	20
Spur-winged plover	21
Dabchick	23
Gulls: black-backed, red-billed black-billed	24
Terns: white-fronted, Caspian	29
Skua	30
White heron	31
Pied oystercatcher	32
Pied stilt	33
Banded dotterel	34
Gannet	36
White-faced heron	37
Henrietta & Thomas: *A love story*	38
Canada goose	48
Coot	49
Bar-tailed godwit	50
Ducks: paradise, shoveler, teal, mallard, scaup	52
Black swan	62
Shags: little, spotted, pied, black, little black	64
Fur seal	67
Harrier hawk	68
Kingfisher	70
Falcon	71
Pukeko	72
Fantail	74
Tui	75
Morepork	76
White-eye	78
Greenfinch	79
Starling	80
Goldfinch	81
Blackbird & Magpie	82
Thrush	83
House sparrow	84
Pheasant	85
Skylark	86
Swallow	87
Whistling tree frog & Stoat	88
Monty the cat	89
Hedgehog	90
Birds in the Waikanae Estuary	91
Acknowledgements	93

KAPITI COAST DISTRICT COUNCIL

Alan Milne, JP Mayor of Kapiti Coast District

Kapiti Coast — a habitat for birdlife

Nearly half of the Kapiti Coast land area is covered in native bush, trees and indigenous vegetation — from Kapiti Island and our coastal shores and estuaries to the hills and ranges in the east, all well connected by river corridors. This is a magnificent habitat for native, introduced and migrant birdlife and the Waikanae River mouth, lagoons and estuary are particularly favoured.

Michael Peryer has become well established on the Kapiti Coast for his personally guided bird tours around the Waikanae River estuary and lagoon areas, attracting visitors from all over the world.

While there are numerous fine publications illustrating our New Zealand birdlife, this new book will appeal to all bird lovers and those who enjoy a good yarn. It will add to the enjoyment of this special locality and will encourage more visitors to this unique area.

Congratulations, Mik, on this fine effort which I am sure will be well received by bird lovers and visitors alike.

Alan Milne.

Mayor, Kapiti Coast District

Waikanae Estuary Scientific Reserve (Waikanae River).
Mik Peryer

Aerial view of Waikanae Estuary Scientific Reserve. The sandspit has been bulldozed to allow the river direct access to the sea. It doesn't take long for it to revert to its natural course and meander southward, recreating the spit between river and sea — with the outlet to the sea creeping ever south.

Charles Lloyd

Waikanae Estuary

Waikanae is one of the best-documented estuaries in New Zealand, with accounts of its birdlife spanning more than a hundred years.

Around thirty years ago development of a marina proposed by the Waikanae Land Company was abandoned. The Waimanu lagoons were formed and there has been extensive settlement of the northern banks of the river. This was expected to have a detrimental effect on the estuary birdlife. This hasn't happened, although some species like bitterns and reef herons have become rarer due to the changed environment. Others have done well and new species have arrived.

Before the Waikanae subdivisions there were salt marshes; now there are two lagoons bisected by a road and surrounded by housing with mown grass verges. Although the estuary is now in an urban environment, birdlife has thrived. It would take little to tip the balance, however — we have to be responsible and look after this jewel at our back door.

In 1946 a Polish diplomat, Kazimierz Wodzicki, conducted a study of the estuary birdlife. Later in 1969–1972 he returned with Max Falconer and Charles Fleming and studied the estuary again, finding that the species and bird numbers hadn't changed a great deal over this period. In 1993–2000 John Bycroft conducted a birdlife study, which also showed a stable number of species.

My observations over the last six years or so indicate that a marked change has occurred since Wodzicki's survey in 1969–72. Numbers have increased dramatically.

Birds in our backyard

We welcome you to our world of birds.

Moira and I have lived beside the Waikanae estuary for ten years. With a lifetime interest in birds, we now enjoy introducing visitors to the birds in our 'backyard'. The estuary is host to over sixty species, including the godwits that migrate to and from Siberia.

This morning the estuary is looking especially beautiful, with the rising sun streaming across the landscape. Kapiti Island is sitting majestically within its wreath of calm sea. It has a mountainous blue tinge, with its valleys and cliffs appearing dark and foreboding. The sea and river add an aquamarine touch, with the painted undergrowth of our estuary reserve contributing soft hues of browns and golden yellows. In the background the hazy curve of the South Island reaches across the horizon like a sheltering arm.

Black-backed gulls are here in their hundreds, dipping and diving, lifting off the water and down again, preening their feathers or just resting on the sandbanks.

A magpie has just flown across from the other side of the river and two large black swans are winging their way upstream.

Mallard and shoveler ducks have been feeding from first light at the river's edge. They are now resting, tucked into the banks with a wary eye ever skyward for hawks, which are away at the moment, probably ranging over farmland. Pied stilts are feeding up the little

Mik, Moira, Meg, Mazie and Minty …

creek, dabbling along the edges. A royal spoonbill rests on one leg at river's edge with its head tucked under one wing. Other spoonbills have left their overnight roost and scattered all over the estuary to feed. Red-billed gulls rest on the sandbank.

A heron flaps quietly along the river verge, reflected in the water. Half a dozen paradise ducks glide in for water after grazing on nearby pasture. Canada geese have turned up this

Waikanae Estuary from our front lawn.
Dominion Post

White-faced heron.
Eileen Thomas

morning, having flown in last night; they will probably lift off shortly. A pair of spur-winged plovers head upstream with their unique weaving flight. Down towards the sea terns are hunkered down in a large group on the beach. Large black shags stand by, along with a family of oystercatchers. These birds tend to keep to this area and not stray much upstream.

The estuary is a beautiful but dangerous place, with its birdlife living and eating according to the rule of nature: the survival of the fittest. The lower reaches are prolific in fish life with the shags feeding on mullet, herrings and abundant smaller fish.

Larger saltwater fish sometimes take a run up the river to see what they can find — at times we have a variety of fish in the estuary, some of which are caught in set-nets. Saw-toothed barracouta now and again range upstream to feed on smaller fish. With their large mouths they are truly fearsome to behold. The river occasionally has paddle-crabs in from the sea to feast on flounder and mullet caught in the set-nets.

During the season people congregate on the banks of the river to fish for whitebait, which run upstream in large numbers to spawn in the swampy upper reaches.

The whitebait run a gauntlet of hundreds of nets each year — the birds know of this event and when the whitebait run, so the diving birds feed. This is when gannets visit the estuary to check on the whitebait. Thus, once a year, the tranquillity of our estuary is disturbed by crowds of whitebaiters.

Upper reaches of the Estuary Reserve.
Colin M McKenzie

Edna Sawtell (85) whitebaiting on the Waikanae River.
Colin M McKenzie

Little shag on the lower reaches of river reserve.
Colin M McKenzie

Upper reaches of the Waikanae River.
Eileen Thomas

Waikanae River.
Colin M McKenzie

The little community

Our estuary is like a little community with the harrier hawk as mayor, councillor, judge and jury all rolled into one. He and his family rule the estuary with an iron fist. If the rest of the birds become disabled, don't stick to the rules, or don't look after their young properly, the rule of the estuary is: *You will be eaten*.

Next in line come the hunters — those that pursue their prey under water, or those that range above the water and dive — all with the one thing in mind: dinner.

Then come the scavengers — the gulls who take their snacks where they can find them and flock in large numbers, especially when food is involved. They squabble and fight for each morsel.

The waders are the elegant members of our estuary community, gracefully feeding along the edges of the water.

Pukeko scratch and peck for their food, both under and above the water on the estuary fringes. Swans with their long necks reach into the water to pluck the succulent plants, but also love to feed on the local pasture. Large royal spoonbills can be found foraging in the mud with their long bills, along with their cousins the small shoveler ducks, who feed in a similar manner.

Calling up and down the river are the spur-winged plovers with the paradise ducks and Canada geese joining in the chorus.

Observing all this activity is our falcon, who swoops down and snatches any bird too slow to get out of his way. When all are in residence together, especially at night, the chorus ranges up and down the scale. The birds, each with their own call, try to outdo one another.

Waikanae Estuary Scientific Reserve.
Eileen Thomas

Eileen Thomas

Royal spoonbill

About twenty royal spoonbills inhabit our estuary, although that number can fall to two or three at times — they commute up and down the coast to Foxton and the Manawatu River estuary. They also nest on Kapiti Island. From our lounge window it's a delight to watch the feeding pattern of these birds as they wave their bills sideways through the muddy bottom of the river. At any one time there may be one or two young birds in the group. These offspring drive their mothers to distraction with their constant chasing and prodding for food, as the adult feeds the young through its bill until they are old enough to feed themselves. The adult bird gets annoyed at times and will fly from one side of the river to the other. Of course the babies follow, so mother doesn't get much respite.

Eileen Thomas

Variable oystercatcher

A pair of variable oystercatchers live on the sandspit, although sometimes there are up to a dozen. The oystercatcher is one of the few seabirds that doesn't have webbed feet; it has toes like a chook.

Our birds were unlucky. They hatched a chick and although signposts were erected stating *Nesting birds — don't disturb*, along came a couple of nosy people and disturbed the birds. It is survival of the fittest in the wild; in swooped a black-backed gull and the chick was dinner.

Spur-winged plover on nest:
note spur on wing.
David Mudge

Spur-winged plover

This rowdy bird lives in large groups and flies with a broken flapping motion. It comes to our estuary and feeds at low tide. Although a creature of the river flats and farmlands, it is territorial and if a hawk gets close it will attack it, using the spur at the elbow of its wing as a weapon. It uses a broken wing ploy to lure you away from its nest. These birds have only been in our estuary for about thirty years in any number, having been blown over from Australia.

Largest and smallest on Waimanu lagoon — swan & dabchick.
Eileen Thomas

Dabchick

The few dabchicks in our estuary can be seen mainly in pairs in the middle of the lagoons, away from the banks. They spend a lot of time under water. It's fun to watch a mother dabchick with its young as she carries them on her back like little bumblebees. The dabchick apparently flies at night — I have never seen one in flight, besides fluttering from one side of the pond to the other if disturbed. Instead of webbed feet, the dabchick has loose webs on its feet and can't walk on land as ducks can.

This dainty bird, the smallest on the lagoon, has a small-person mentality — it is not going to be bullied by anyone. I was observing a beautiful little dabchick on our lagoon minding his own business, when he became hemmed in by some cheeky mallard ducks feeding. He stretched out his neck and paddled flat out at a large mallard drake and pecked its back. The mallard beat a hasty retreat. Our dabchick then sat up in the water, shook himself and flapped his wings as if to say 'I may be small but don't meddle with me'.

Dabchicks.
Geoff Moon, DoC

The gulls

A receding tide left a puddle which trapped a large number of herrings. Every bird in the estuary told their mates about this good fortune and they came in their hundreds. The terns dived for the live herrings and the waders, pied stilts and herons all had their fill. As the water evaporated the fish died and the scavengers — black-backed gulls — cleaned up the beach. In a little while everything was back to normal.

Eileen Thomas

Gulls with Kapiti Island in the background.
Colin M McKenzie

Black-backed gull

The black-backed gull is a scavenger and will eat anything it can see on the beach, often lifting a pipi into the air and dropping it from high to crack the shell. When the southerly blows, all the gulls fly down our side of the river to get lift from the escarpment. It's impressive to watch them soar against the wind without flapping their wings.

There are hundreds of these birds in the estuary and river reserve. It is easy to tell young from adult: although they are of a similar size, the young birds are dark brown and are usually chasing mother for food.

These large, powerful birds are common inland around larger river systems. They have taken a liking to lamb meat and unfortunately kill weak lambs and cast ewes. They are also useful scavengers, however, cleaning up dead animals.

On Waikanae Beach.
Eileen Thomas

Red-billed gull

There is quite a population of red-billed gulls in the estuary. They fly up the river in groups of about twenty and can be found with the Caspian terns on the sandspit. It was one of these gulls that our harrier hawk played 'bar the door' with and took out (see page 68). I suspect that the hawk had young to feed and the hapless bird ended up as gull pie.

Clearing my fishing net in the river one day I noticed a red-billed gull flapping unnaturally in the middle of the river — a nylon trace from a fishing line was wound around its body. The flapping and the current took it into a willow branch which had washed down in the last flood. The bird was against the other bank so I swam across the river to free it. I looked up to see a stoat standing on its hind legs to see what the flapping was all about. The stoat apparently thought to get itself a cheap meal. I managed with a bit of difficulty to unwind the nylon and free the bird, throwing it into the sky — off it flew, lucky bird.

Peter Morrison, DoC

Black-billed gull

Similar in size to red-billed gulls, these birds live mainly on South Island rivers and feed differently. They are in our estuary in small groups, distinguished by black feet and slim black beaks.

Rod Morris, DoC

Caspian tern

This handsome bird may at times be seen on the sandspit or lower river verges. A large, beautiful bird, it has a bright orange bill and black topknot.

Rod Morris, DoC

White-fronted tern

These distinctive birds hunker down in groups on the sandspit, diving for food in the sea or upstream within the estuary reserve. Their beautifully curved wings and black topknot create a streamlined look. They come and go. Sometimes they are there — next thing they're gone.

Eileen Thomas

Skua

Skua visit the estuary at times, and it is hard to distinguish them from similarly coloured young black-backed gulls. Skua often chase other birds to make them disgorge fish they have eaten.

DoC

Don Merton, DoC

White heron

The white heron is a rare visitor to the estuary. It is a distinctive large bird and is easy to differentiate from the royal spoonbill, which is about the same size and colouring. The heron flies lazily with its head pulled in, whereas the spoonbill has its beak and neck extended and flies with purpose.

Peter J McClelland, DoC

Pied oystercatcher

One of our oystercatchers has a broken leg and has to hop along. In the wild this is serious — it is doubtful that this bird will survive for long, although it has been injured for a few months now. Oystercatchers forage on the beach and tend to be fast on their feet, so it is at a disadvantage.

Pied oystercatcher with godwits.
Dick Veitch, DoC

Pied stilt

The pied stilt is a truly lovely little bird, elegant on its long stilt legs.

In the twilight right on dark it is ghostly in flight, with its black and white colouring. Stilts have a sharp bark-like cry and are found mostly on the river verges or the sandspit.

Pied stilt on the Waikanae River.
Eileen Thomas

Banded dotterel

It is a pleasure to see the banded dotterel's plumage, especially in the breeding season. The bright orange band across the chest is a sure giveaway. Dotterel are friendly and not at all afraid of humans. They are remarkable as one of the few birds to commute to Australia against the prevailing wind.

Only little, the birds have hooked wings like a swallow. They have a short, sharp feeding run and like the verges of the lower river or the sandspit. As they blend into the surroundings they are hard to see.

Peter Morrison, DoC

David Mudge

David Mudge

Gannet

These birds usually feed out to sea or fly along outside the breaker line of the beach and dive from a great height for food, usually singly or in pairs. When the tide is full they make their way upstream to wheel and dive for food in our reserve.

White-faced heron

In the evening herons flap slowly past our windows to roost in the pines of Otaihanga. At times, with the sunset glowing yellow in the background, it looks like a Chinese landscape painting — dark birds on a light canvas.

Our lagoons are bisected by a roadway; a large concrete pipe connects them under the road. The herons were quick to catch on that if they stood on the pipe instead of hunting for their tucker around the edges of the lagoon, their food would come to them. You may find a heron standing like a statue on top of the pipe, waiting for its dinner to swim into view. One large heron thinks the top of the pipe is his — and when he's had his fill, flies into the pine at water's edge to rest. If any other heron spots the pipe free and flies over for a feed, this large heron will dive-bomb it, chasing it away to protect its patch or rather its pipe.

Fishing from the pipe.
Colin M McKenzie

Waimanu Lagoon.
Eileen Thomas

Henrietta and Thomas: a love story

Henrietta, our injured black swan, is at least eighteen years old and probably nearer twenty, so she hasn't long to live. Thomas the goose has been her mate for some thirteen years. Their relationship is stable, although a few years ago a gaggle of about thirty geese moved from the river to the lagoon where Thomas and Henrietta lived. The urge was too strong for poor old Thomas and he took a mate of his own kind, but this didn't stop him keeping an eye on Henrietta.

Rex, a neighbour, was walking his corgi along the grass verge of the lagoon when Thomas — neck outstretched, wings flapping, hooting raucously — charged poor Megan the corgi, grabbed her by the foot and wouldn't let go. Megan was most upset. She had never been attacked by a goose before and her squealing echoed around the pond. Rex had to rescue his dog but was rather scared himself, so they quickly retreated.

The local rangers decided in their wisdom that while it was okay for geese to live on the river, it was too much for them to pollute the grass verges of the lagoon. The order was out — *Get rid of the geese*. Now the rangers didn't know which goose was poor old Thomas and no one wanted him to go, as Henrietta needed him. So Ethel, who lives by the lagoon, sprayed Thomas with bright red food dye. He was saved — in the dead of night the rangers pounced and the geese were gone.

A pair of black swans made a large round nest beside the lagoon and raised seven cygnets. They were proud of their brood and swam on the pond with the cygnets between them.

The lovebirds.
Colin M McKenzie

Mother would take the cygnets up the pond to a secluded area and spend most of the day there. In the evenings, for the first few days, she would bring them back to the nest for the night. The cob, father swan, would stay near the nest during the day and chase other swans away from his area, stand on the bank and stretch out his neck. He'd have a good old flap of his wings to let everyone know 'This is my patch!'

Now Thomas didn't like this swan chasing his Henrietta, so he climbed onto the bank, charged at father swan and beat him up. He grabbed him by the base of his neck and, flapping his wings, let him know that he was not having anyone chasing his mate. Then Thomas stretched out his neck, flapped his wings, swanked his swank — he was boss around here.

Henrietta and Thomas had been together for more than a decade when a second swan entered the equation. Henrietta fancied this new bird, and so an uneasy love triangle was established, with Thomas protecting both Henrietta and the recent arrival.

Everything changed when this new black swan laid an egg.

Colin M McKenzie

The new swan was a female! No one expected the egg to be fertile — after all, Henrietta was female. Even so, Henrietta helped make a large round nest where our new swan laid another four eggs. Both began to sit on the nest.

Thomas wasn't too happy with this new arrangement. He was losing control and the swans were doing their own thing, so he started harassing them. The swans have quite a ritual when they change partners on the nest. They talk to each other and raise and lower their necks, then one moves off and the other moves on.

This is when Thomas would chase them and beat them up. They became terrified, especially the young female. Thomas would squawk loudly, flap his wings, stretch out his neck and chase them for some time, pecking them if he could. This caused great difficulty when the birds changed partners and the swans broke two of the eggs. When the change took place and one of the swans moved away Thomas would settle happily next to the one on the nest.

It's hard to establish the gender of mature swans, so it was with some surprise when the three eggs hatched — establishing that our 'Henrietta' was in fact a Henry.

Now that everyone has moved off the nest Thomas is happy again and protects the whole family with the vigour he used to protect 'Henrietta'. I think he is rather proud of his step-cygnets.

The family — Thomas, 'Henrietta', the new swan and the three cygnets — are famous on our coast and beyond, as it is rare to find two different species cohabiting in the wild.

As the cygnets grew up we wondered how they would react to their parents and step-parents' inability to fly away with the other swans.

Colin M McKenzie

Changing partners on the nest.
Eileen Thomas

Thomas took it upon himself to teach them to fly. When they had matured and had flight feathers he would chase them with much squawking around the lagoon until they lifted off and landed again. This game continued until they could lift off without help from Thomas. When they decided to try flying themselves, Thomas would stand up in the water, flap his wings and cry his loudest to encourage them.

Now that the young are mature birds, Thomas and 'Henrietta' have chased them off the lagoon. They are happy to be back to the eternal triangle with the female swan.

Another pair of swans had a late hatching of six cygnets on the river upstream from the lagoons. They took their brood across the river and downstream to the roadway that separates the river from the first lagoon. They crossed this roadway into the first lagoon, then proceeded to the other end where another roadway separates the first lagoon from the second.

They obviously had plans to raise their family in the seclusion this lagoon provides, but they didn't reckon on Thomas and Henrietta. After chasing their brood off their lagoon the couple certainly weren't going to allow interlopers to disrupt their set-up.

Thomas and Henrietta met these two swans, with their seven babies, squarely in the middle of the roadway. With wings flapping and much racket Thomas and Henrietta chased them back into the first lagoon.

I wonder what the cygnets thought of that?

Just out of the egg.
Colin M McKenzie

Eileen Thomas

Would you believe it? 'Henrietta' (Henry) and the new swan have been at it again, making up for lost time. They built a lovely new nest in the most exposed site on the lagoon and laid seven eggs right next to the little footbridge where everyone walks, including dogs and children. The swans don't mind this influx, as they get a few extras in the form of tidbits. The dogs are a worry, although the powers that be have erected on high poles signs stating *Dogs on leads*. I have supplemented these with my own signs at a sensible height: *Nesting swans — secure your dog.*

Thomas is pleased now he knows what his swans are up to. He is content to stay on guard and sit out the incubation.

Stop Press

Those swans are still doing their thing — Thomas has six more step-cygnets.

'Henrietta' is now called Henry after it was established that 'she' was a male. The new swan has taken Henrietta's name and Thomas is still Thomas. With the hatching of the six cygnets everything settled down — Thomas was in charge and fussed around looking after everyone again. Just before the cygnets had their flight feathers, the mature swans forcefully chased them away onto the first lagoon. The poor little cygnets couldn't make out why mum and dad had gone from being so loving to being so aggressive, especially as they were still young and hadn't yet learned to fly.

Henry and Thomas decided to make up for all the lost time of those thirteen gay bachelor years when there wasn't a female swan to mate with. The third nest was started in record time — this is why the two swans were so aggressive towards their previous brood.

Although Henry, being the male swan, mated with the female, Thomas the goose had his turn as well. I have it on good authority that a goose can't fertilise a swan, but time will tell!

Subsequently the swans made a new nest in a secluded area of the lagoon and laid seven more eggs. As they established their nest they both became aggressive towards poor old Thomas and gave him a hiding, knocking the stuffing out of him.

During the incubation he generally kept his distance and spent most of the time across the narrow area of water on the far bank, for the 40-odd days it took to hatch the brood. He would swim over to the nest to check it out at times and was still vigilant about strangers and dogs.

While sitting on their eggs swans are brave — they won't come off the nest if anyone approaches. They hiss and honk, fluff up their wings and increase the size of their necks dramatically, then undulate their necks back and forward close to the ground, for all the world like a snake.

This fierce display is usually enough to scare off most pests, but would you believe it, two young schoolkids attacked our Henry on his nest with sticks. Rex, the fellow

First outing.
Mik Peryer

Look — I have wings.
Eileen Thomas

Henrietta chasing the cygnet who can't yet fly off the lagoon.
Colin M McKenzie

whose corgi was bitten by the goose, saw what they were up to and chased them off before they did any damage.

This time all the eggs hatched and so the continuing love story has yet another chapter. About eight weeks later our cygnets started to lose their baby fluff and slowly changed colour with darker feathers showing through. As they grew they were taken waddling across the roadway to the southern lagoon — to explore and to be shown off to all the other inhabitants — where Henry spent most of the time chasing his previous brood away from the new babies.

While they were minding their own business on the southern lagoon, along came some pesky humans with kids in tow. They launched a large orange rubber dingy with a smoke-belching noisy motor and whizzed around the lagoon dropping buoys. Then little sailing boats appeared with kids in them who proceeded to sail all over the place.

This was too much for our family. Mother swan (the only one who could fly any distance) took off for the river, leaving Thomas and Henry to round up the cygnets and high-tail it out of there. Thomas, calling loudly, led the way as fast as his little legs could carry him. Our babies had to run to keep up and Henry plodded along behind. Along the bank they waddled, then across the roadway — thank goodness there weren't any cars coming — to the seclusion of their own lagoon.

Henrietta turning over the eggs.
Eileen Thomas

Thomas chasing the ducks away from the cygnets.
Eileen Thomas

Along the bank.
Eileen Thomas

Thank goodness there weren't any cars coming.
Eileen Thomas

To the seclusion of their own lagoon.
Eileen Thomas

In the corner of the Waimanu lagoon — over the bridge where the swan that Thomas beat up made its nest — are two gravestones. One belongs to Margaret Maria Durie who died in 1848, the three-year-old daughter of Waikanae pioneer Major Durie. He has a street named after him nearby. The other gravestone is of a whaler, William Browne, who drowned in 1852 off Kapiti.

Cygnet, four months old.
Eileen Thomas

Canada goose

A formation of about fifty Canada geese fly into the estuary at times, always late at night. They zoom in over our sandhill and, chattering and squawking, swoop down into the river to hoot and squabble all evening. In the morning they form into a squadron and away they go again.

We have two resident Canadas; I suspect they have mated up and made the estuary their home.

Coot

This is another recent rare visitor, although coots have been in the Horowhenua area for a while. They can't be silly coots but must be smart coots to be the first of their species to arrive in such a wonderful place.

Colin M McKenzie

Bar-tailed godwit

About twenty godwits visited our estuary some years ago, but this figure has dwindled to about seven this year. They appear October/November — sometimes in groups, sometimes in pairs. They stay for a week or two then move on.

We have a pair of godwits wintering over. Having decided not to migrate this year, they are staying here in Waikanae, probably until next year when they may, with the majority of godwits, make that great flight back to their nesting area in Siberia.

One of the pair has a broken leg. It is completely useless, swinging like a pendulum from his body. It's sad to see the bird like this as he won't last long.

He manages well, prodding the mud for his food while standing on one foot, using his flapping wings to balance, then hopping along feeding on the edges of the river. He gets no help from his injured leg.

He's been on the estuary for four weeks now. Let's hope he comes right and survives long enough to get back to his breeding site on the other side of the world.

Will the hawks get him? Not at the moment, as he is still in good order, able to feed himself on the abundant food and get around with that swift flight of his.

Bar-tailed godwit, on a visit to the Manawatu.
Alex L Scott

When the food supply diminishes, as it does from time to time, he may find he can't feed from the edges of the estuary and hold his own with the other birds. He could lose condition, and with his disability, perhaps the predators will get him.

Godwits migrate from Alaska and Siberia — they have travelled a long way. Their young, who are only a few weeks old, accompany them. They wait for a big gale with 100–150 kph winds from the north, then away they go on the longest nonstop over-water migration of any bird, across the widest part of the Pacific. It's about 6000 kilometres to New Zealand and the Waikanae estuary, taking six days and nights at speeds of up to 80 kph. What a truly amazing feat of flying!

Wouldn't it be a tragedy if, when they got here, we hadn't looked after their feeding ground — the Waikanae estuary — as we should?

Bar-tailed godwit with injured foot on Waikanae River.
Colin M McKenzie

Our ducks

With motley brown feathers ducks blend in splendidly with the undergrowth, and when nesting sit very still on their eggs if danger is about. While walking along the rank grass verges of the river looking to cast my fly line into a likely pool I nearly jumped out of my skin when a hen bird exploded into the air from under my feet, straight off its nest which contained ten eggs.

On our lagoon we have a Pekin duck. Pekins are white — have you ever seen a white duck trying to blend into the undergrowth while sitting on eggs? As you approach she quietly makes herself smaller by lowering her head onto her body, but she still sticks out like a sore thumb. She is brave — she won't come off her eggs. I wonder if she will manage to hatch them, as she is so conspicuous.

Pekin ducks are bred for the table as they make good eating. A farmer friend bought two dozen Pekins and penned them in near a creek, with a run over the water so they could have a swim. Being enclosed, they were safe. The creek flooded, however, and the pen washed away. The ducks were left to roam the farm, always keeping together.

On this farm lived a harrier hawk and his family. Harrier hawks love ducks, and seeing the Pekins in the paddocks they pounced, eating four ducks over a week or two. This upset the rest of the ducks so they devised a way to survive. Pekin ducks are smart — they changed their habits and during the day waddled in line across the road and into the native

Colin M McKenzie

bush reserve on the hillside. They spent the daylight hours hidden in the scrub and during the evening when the hawks were roosting, waddled back to the creek and fields to feed. They actually changed their habits and became nocturnal birds to survive the hawks.

A Muscovy duck also lives on our lagoon. She is quite large, dark in colour, and has a white sash down her chest. She had a hatch of thirteen ducklings. However, spring is a wonderful time for our hawks, seagulls and pukeko. They love the ducklings, as do the large eels that lurk in the estuary ponds. It took just one week for Mrs Muscovy to lose all her brood. There were some contented eels, hawks and seagulls about. The eels don't have it all their own way — large black shags scour the lagoons for them and when they catch one it's a fight to gobble it down before it wriggles out.

Help!

My farmer friend Collis had a dog named Joe who loved to chase the wild ducks down on the creek. Collis was forever calling him to heel. One day he didn't notice Joe was missing, until a mature mallard — quacking loudly — flew at him across the field, straight and low. Agitated, it circled around Collis then flew back to the creek where Joe was chasing its ducklings.

Obviously, it needed urgent help — it sure knew where to find it. An amazing interplay between wild bird, farmer and dog.

Bottoms up!
Colin M McKenzie

Paradise duck

Between the shingle banks in the river, the grassed paddocks of Otaihanga and further afield roams a large flock of juvenile paradise ducks. They like to come back to the river for water. Other places are occupied by the older mated birds who have kicked out the young to fend for themselves.

The male looks dark from a distance but is in fact brightly coloured. The female is the one with the white head. Paradise ducks have been known to nest high up in large trees or use high-tension power poles as perches.

On farms, dogs spend ages chasing 'parries' off the paddocks. The ducks use minimal effort to evade the dogs, lifting off and landing again.

Shoveler

Because of their feeding habits these ducks appear shy. They don't congregate around for food as other birds do, but use their beaks to sieve the water and mud for food. If you look out to the middle of the pond you may see the brightly coloured male and his less vivid mate, both looking top-heavy with their large spoon-like bills. The shoveler is also known as a spoonbill duck or spoonie.

On Pukepuke lagoon at Foxton in early February a shoveler duck was banded when only a few days old and a nasal saddle inserted in its bill to make identification easy. This bird was seen in Southland several times in May. By September it was back on the Pukepuke lagoon where it raised seven ducklings. It was not seen again until next breeding season when it came back and raised another brood. Then it was seen no more.

A mallard with a brood of fifteen two-day-old ducklings was edging along the undergrowth at the side of the lagoon when a pukeko grabbed a duckling. Our shoveler saw what was going on and — quick as a flash — flew over and chased the pukeko away, rescuing the duckling. Another fascinating inter-species drama.

Eileen Thomas

Grey teal.
Mike Aviss, DoC

Teal

Chestnut

The five chestnut teal in our estuary have only recently arrived from Australia. According to the Department of Conservation they are a rare visitor and may have flown over to escape a drought. It could be out of the frying pan into the fire, as we can have drought on the Kapiti Coast, and the duck-shooting season has started. But they will be safe as they are a fully protected species and there is plenty of food and water here.

It is especially pleasing to see rare species in the estuary reserve every now and then.

Grey

Grey teal are beautifully compact birds which fly rather like a starling, in a close weaving formation. These birds are fully protected and are often seen on Waimeha lagoon.

Brown

If you look in the carpark area at the road end by the river, or just upstream, you may find four brown teal minding their own business away from those pesky mallards. If you disturb them they may just paddle quietly into the river out of harm's way.

What a beauty: a brown teal on Waimanu lagoon.
Eileen Thomas

Mallard

British settlers introduced the mallard as a game bird. It reminded them of the countryside and sport of the old country — and supplemented the larder.

Mallard are adaptable. When the weather is fine and the shooting season is upon them they fly *en masse* to raft up out at sea. They sit out the daylight hours and fly back for dinner in the evenings, thus surviving the shooting season. The older birds tend to live for five or more years — they are cunning. When coming in to feed in a promising pond they send the younger birds in first, so the immature birds succumb to the hunter and the wise heads live to see another season.

In the lower North Island ducks follow the harvest from field to field and farm to farm. They work their way down the island, feeding on barley and grain left over from the harvest.

They also love to eat peas grown for seed. In their pods the peas are safe until ready to harvest. When ripe, however, the ducks waddle across the crop, popping the peas out of their pods. They arrive in their hundreds, sometimes thousands. They quack-mail over large areas to their mates, announcing when the peas are ready. The birds land in the middle of the crop and work their way out in a circle. You can see where they have trampled the crop down flat. It looks like a spaceship has landed, with large rings in the middle of the field.

On the lagoons, mallard, grey ducks, white Pekins and even the odd Muscovy have loose morals — they breed with all and sundry. As a consequence, it's nearly impossible to tell different breeds apart.

... with baby on mum's back, Waimanu lagoon.
Colin M McKenzie

Scaup

We have a family of scaup who come and go. Sometimes there are dozens of them — then they are gone. Dark in colour with the males having bright yellow eyes and the females brown, they are divers and spend quite a time feeding under water.

Scaup have increased dramatically on our Waikanae lagoons, from only half a dozen a few years ago to a group of about thirty who now breed here.

Eileen Thomas

Water-skiing.
Colin M McKenzie

Black swan

The swans commute between Cape Farewell in the South Island, the Wairarapa lakes, the Manawatu and further afield. They are expert flyers, travelling high and fast.

Our black swans, who raised the seven cygnets, have had a hard time of it lately. The cob that Thomas the goose beat up was found dead on the lagoon one morning. It hasn't been established how he died — perhaps it was naturally. It was a sad time, as Mrs Swan was left a solo mum. At this stage, the cygnets were large and mother didn't have too much trouble raising them herself. They have subsequently fledged — we see them flying upstream, gathering strength and growing daily. Back on the lagoon, however, it's another story.

At times the adult swans can't tolerate these younger birds coming back. With necks outstretched and much flapping of wings on water they will chase them across the lagoon. If they can, they'll peck them on the behind. Then, pleased with themselves, they will go about the business of being contented swans again. The young birds are sometimes so afraid that they will clamber up the banks to get out of the way of these cranky older birds.

Black swans in flight are majestic, with necks outstretched and their white wingtips contrasting with their black body like flags.

At times when coming in to land, especially after a long flight, they will crook their wings and fly almost straight down, then water-ski to a halt if landing on water. Black swans also land on pasture, as they are grazers and love feeding on farmers' paddocks.

See my wings!
Colin M McKenzie

Little shag

The smallest shag on the estuary has sleek black plumage and a white face and throat. At times it may have a white waistcoat. A solitary bird, it prefers to sit on its own in the cabbage tree at the southern end of Waimanu lagoon, or on the pines at the perimeters of the lagoons, when not fishing. Little shags prefer the river and freshwater lagoons and don't go out to sea as other shags do.

Colin M McKenzie

Spotted shag

This fellow has yellow feet, a speckled grey body and wings, twin stripes down his neck, and two crests on his head. Our visitor had a silver band on one leg and had been banded on Somes Island in Wellington Harbour, so hadn't travelled far…

Peter Morrison, DoC

Pied shag

Found at times on the pines around the lagoons, this handsome bird has a distinctive black and white body.

Peter Morrison, DoC

Black shag

Black shags can be seen on the sandspit or a log with their wings outspread. They need to dry their wings thus because they lack the oil that other birds have. This enables them to swim underwater, but they tend to get waterlogged.

A net set in the river was raided by a black shag whose eyes were larger than his belly. He spied a nice-sized flounder, managed to free it, took it to the mudflats and tried to swallow it. Now a flounder is too large for any black shag, and he had to abandon it. A black-backed gull swooped down, lifted up the flounder and flapped to the far bank where my friend Jeff happened to be. Jeff rather fancied a flounder so he shooed the gull away and it dropped the fish … Jeff had hand-me-down flounder for dinner.

JL Jendrick, DoC

Little black shag

The little black shag is the vandal of the estuary, with up to thirty flying and fishing together. What a prodigious larder the estuary is! When the tide goes out and the sandbars are exposed, little black shags fish the river in squads. First they peel off from the front to dive. Then, when overtaken by the rear birds, they pop up and fly to the front again. They trap herrings and other fish against the shingle bars where there is a frenzy of feeding.

Herons on the river also know what's going on. They fly over, hop along the sand, and snatch the little fish which throw themselves out of the water to escape the shags.

Drying their wings.
Colin M McKenzie

Fur seal

It's a surprise to walk the dog along the river bank and come across a baby seal. A hasty putting on of the dog's lead and next thing, "oink oink" — the seal's into the river and on its back, slowly waving its flippers. Then back to sea again.

I had a phonecall from my neighbour one wet afternoon. He had caught a seal in his set-net in the river and dragged it onto the sandbank but needed a hand to free it. I grabbed a sack and scissors, put on my chest waders and off to the estuary I went. After putting the sack over the seal's head we snipped up the net. Seals are very smooth and out it popped. I was standing between the seal and the water and mister seal didn't think that was right. He came at me and, as I can't run fast backwards with waders on, bit me on the knee of my waders, which luckily were reinforced. He didn't do any damage. Feeling no doubt pleased with himself, he flapped his way into the river and was gone.

The seal that bit me, on Waikanae River.
Mik Peryer

Harrier hawk

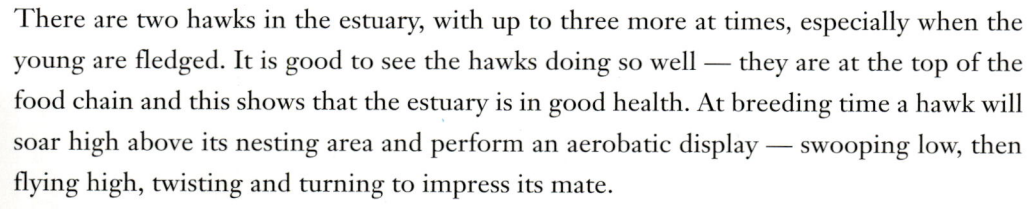

There are two hawks in the estuary, with up to three more at times, especially when the young are fledged. It is good to see the hawks doing so well — they are at the top of the food chain and this shows that the estuary is in good health. At breeding time a hawk will soar high above its nesting area and perform an aerobatic display — swooping low, then flying high, twisting and turning to impress its mate.

One windy day I saw a hawk trying to stop a seagull getting back to the sea. The seagull fled upstream then turned again to come back, and the hawk took it out of the sky. This is rare, as harrier hawks are normally inclined to snaffle smaller birds and mice, or to scavenge. However, I have seen a hawk worry a flock of starlings until they became confused and it was able to catch one of them on the wing.

Magpies, plovers and sometimes swallows will dive-bomb hawks if they get too close to their space. When the southerly blows the hawks hunt on our side of the river, using uplift from the escarpment to propel them along. We can look down on them and almost reach out and touch them at times.

When birds resting on the shingle banks are disturbed, if it's not dogs or people, you can bet it's a hawk. They will grab a duck, given the chance. When out duck hunting, I had a harrier hawk swoop in and thump onto one of my decoys. He must have had a hell of a fright!

Our resident harriers like to vary their diet. Sometimes they fly directly to Kapiti Island, one of our most important nature reserves, to feast upon whatever they can catch. It doesn't worry them one little bit if the bird they are scoffing is endangered.

David Mudge

Our farmer friend decided to trap some magpies and baited a netting trap with a hunk of venison. When the trap was checked, lo and behold, it had caught a harrier hawk which had eaten all the venison. When the trap was lifted to let the hawk out, it was so full that it couldn't fly, and had to waddle away across the paddock.

Rod Morris, DoC

Young kingfisher.
Rod Morris, DoC

Feeding young.
David Mudge

Kingfisher

Several pairs of kingfisher live in the estuary and around the freshwater lagoons. They are popular with humans but renowned for aggressive behaviour towards other birds. One was seen killing a fantail, swinging it around in its beak and banging it against a tree branch. Perhaps it had a troubled childhood.

One day I went to visit my neighbour who, as it happened, wasn't home. I rounded the corner to his back door and spotted a magnificently coloured young kingfisher on the concrete. It saw my dog and me, fluffed up fiercely and shrieked with its beak extended. I managed to control my dog and catch the poor thing before my dog did. I thought it may have been injured, but it didn't appear to be. Maybe it had flown into a window and was disoriented. I took it to our local bird sanctuary and the lady there checked it out. She fed it a small piece of meat, poking it down with her little finger. Next day I rang and it was doing fine.

As its name implies, it is an excellent fisher, but spends a lot of time catching insects. It likes the fences around deer farms — the extra height of the deer fence makes a handy vantage point. It will use this to work right around the paddocks, catching food among the grasses.

Falcon

Falcons are occasional visitors to our estuary. One day I heard a thump as a falcon snatched a starling out of the air and slammed into the ground with it. Another flew along the escarpment and landed on a pine tree. It looked magnificent.

Falcons can be aggressive, and have been known to attack people while protecting their territory. A fellow who had invaded their patch was dive-bombed by a falcon. It lifted off his hat and swooped into the sky with it, then released it. Its mate caught it on the way down.

The other evening a falcon scooted at top speed over our heads, dead silent, with no wing noise at all. It swooped down over the river and was gone.

My farmer friend had a falcon which had made its home in the native bush around the farmhouse. In the same area lived scores of sparrows who liked to feed on the grain and barley seeds from the silos and haysheds.

The falcon rather liked this set-up as he could sit high on a nearby power pole, preen himself and when he felt peckish, dive into the barn and help himself to a sparrow. He would then fly back to his power pole perch, have a sparrow dinner and watch his next feed getting plumper on the farmer's grain.

New Zealand falcon.
Rod Morris, DoC

Pukeko

The many pukeko in the estuary like to range over the grassland and now and again fly to our side of the river.

Pukeko are well educated. Those living near the roadway cross over to entice the humans to fork out some tucker. Then they stop on the roadside to make sure there's no traffic coming, looking again before crossing back into the estuary.

They are an ungainly bird, always appearing awkward, but it's amazing how they get about. They nearly always fly at night, covering long distances.

Two birds were banded in the same area. One was found 90 kilometres north four months later and another 150 kilometres south.

Pukeko are a worry to other birds as they will take babies out of nests and even take ducklings if they get the chance. They can be cranky if other birds invade their patch. I saw five oystercatchers fly into the little creek on the river where the pukeko like to forage. Straight away, two pukes chased the oystercatchers away. This isn't a usual haunt of the oystercatcher, as they prefer the beach and lower reaches of the river.

Pukeko are territorial. Two groups were observed over a period with one group living downstream from a clearing in the bush, the other upstream. At times they would meet at this clearing. It was almost as if there was a straight line down the middle of the clearing, with one group daring the other group to cross. If they did venture across this imaginary line all hell let loose and it was all on with much screeching, running and jumping until one or the other group retreated and all was tranquil again.

Rod Morris, DoC

Fantail

The fantail is a deft bird of the bush and surrounds. It uses its attractive tail to change direction when flitting around chasing insects.

Fantails appear friendly, with little fear of people, especially when you're walking through the bush. This isn't because they are pleased to see you — rather, they are pleased to see you disturb the insects they feed on. Our fantail darts from bush to bush on the escarpment and now and again, when insects are about, will come into our house — flying around the ceiling and disrupting our household no end.

Our two cats wake from their slumber. With eyes like saucers, they snap into hunting mode with one thing on their mind: fantail pie. Meg, my dog, doesn't like this intrusion one bit and will bound about barking. Pandemonium reigns. There's real relief when the fantail flees out the open door and things return to normal.

Fantails on the farm are quick to catch on when food is about. At least twenty wait for the dairy cows to finish milking, following them back to the paddocks. While grazing the new pasture, cows disturb the insects that dine on their droppings and the fantails feast on the insects — the cows do all the work.

Fantails have even been known to land on fishermen's rods as they angle for trout.

David Mudge

David Mudge

Tui

My dog Meg has claimed the top of the sandhill as hers. One day she barked, growled and ran back to tell me something strange was in the bushes on our escarpment. I investigated and what did I find but three tui seated on a flax bush. It is unusual to find tui in our estuary but they love flax in flower. Meg, a gun dog, knew they were birds she didn't recognise. She has never worried about the other birds — black-backed gulls, smaller red-billed gulls or all the little finches, sparrows and swallows that are about — she just knew these were different.

The tui, with iridescent colouring and white cravat, is a truly magnificent bird. It is often referred to as the bully of the bush as it is possessive, and will protect its patch with passion.

Helen Gummer, DoC

Morepork

A resident morepork (owl) lives on the fringe of the estuary just upstream from the edge of the scientific reserve. On a clear frosty evening its call can be heard everywhere. Although not regarded as a bird of the estuary, because of where it lives I have no doubt that it ranges throughout the area. Right at dusk, I have had a morepork land on a bush I was standing under. I don't know who got the bigger fright.

The owl was back in the bush happily roosting in a shady nook. All the birds in the vicinity told their mates it was there and they ganged up on it. Three cheeky fantails arrived along with five blackbirds and voiced their disapproval in no small manner. The fantails chirped angrily, flitting around being brave. The blackbirds clucked their warning call, flying from branch to branch, knowing that the owl can't see well during the day. It ignored them all, but they were careful to keep their distance. It all changes in the evening. During the day the birds can be brazen because of the morepork's eyesight, but as night falls it becomes the predator.

Our floor-to-ceiling windows overlook the estuary and we don't draw the blinds. Often there are moths about and our security lights are a magnet for these and other insects. I heard a thump and went to investigate — nothing. Later there was another thump and I saw a little morepork on our bird-feeder. He was chomping moths and the lights must have blinded him — he had hit the glass. Fortunately it didn't hurt. He flew off into the night, landing for a time on my neighbour's sun umbrella. It was the first time we had seen a morepork in the estuary proper, although I always knew they ranged over the area.

Dick Veitch, DoC

Baby morepork.
CD Roderick

Morepork country, Waikanae River.
Eileen Thomas

White-eye

The tiny green white-eye has arrived, staying only a while before moving on. Flitting from branch to branch, it is a lovely bird with a distinctive white ring around its eyes. Sometimes it's called a silver-eye or wax-eye.

White-eye.
Rod Morris, DoC

White-eye.
Peter Reese, DoC

Greenfinch

The sleek greenfinch feeds on our lawn on top of the sandhill. It mixes with the sparrows and blends in with them, so you have to look hard to distinguish them.

Greenfinch.
David Mudge

Starling.
Dick Veitch, DoC

David Mudge

Starling

Starlings must have a clock installed because at certain times of the year they gather in large flocks in the evenings, flying over the estuary in their hundreds to roost on Tokomapuna Island between the mainland and Kapiti, almost as if ready to migrate. Starlings love to sit on the backs of sheep in the paddocks, where they have a vantage point to scan the grass for insects.

Goldfinch

On a lavender bush on the sandhill we have had about fifty goldfinches all feeding on the seedheads. It's amazing how these beautiful multicoloured birds fly in unison, tightly bunched. They come and they go — sometimes there are none to be seen then they turn up again.

Blackbird

A friendly blackbird with a white feather in one wing lives at the edge of the lagoon on the corner of Tutere Street and Barrett Drive. It's been around for three or four years now and nests in the same area every year. Pied blackbirds are not uncommon — I have seen several in other areas. The female blackbird isn't black, rather a dull brown.

Our blackbird.
Eileen Thomas

Magpie

Magpies visit regularly and feed off grass grubs on our sandhill. Our dog Meg isn't having any magpies invading her territory and chases them off.

The magpie is an aggressive killer of small birds and is fiercely territorial during the breeding season — it wreaks havoc on nests, eating eggs and young birds. A magpie will harass roving harrier hawks and chase them off. Keep well away from its nest or you'll be dive-bombed.

DoC

Thrush

Mother thrush had a red berry. We weren't sure what the berry was, but it may have been from a fuchsia. She fed it to her baby, who promptly spat it out. Mother thrush caught it and put it back into baby's beak and it was spat out again, to be caught a second time in the air. Mother then peeled the skin off and again gave the berry to baby. Baby spat it out again, only to be caught once more. Next, mother deposited it right down baby's throat. It had to be swallowed. There seems to be a parallel to human behaviour here.

A nest had just been vacated by a family of thrushes and our friend Kay put her hand into the nest to see how warm it was. What happened? She was covered in mites who were looking for another host.

If they have a vantage point such as a tree, thrushes can spot worms in a lawn from over ten metres away. This almost inexplicable feat has been observed many times.

A thrush nested and settled down to lay her eggs, when a pair of brazen starlings came and stole her nest, shooing her away. Starlings are untidy and when they nest they usually do it on something flat. They love to get under house roofs, making a mess on the flat ceiling. When these starlings tried to make the thrush's nest to their liking, most of the nesting material ended up on the ground. It was never retrieved — they preferred to find new material. After a while they gave up and our thrush was able to reclaim her nest. She proceeded to lay her eggs and raise her family once more.

David Mudge

Chris Smuts-Kennedy, DoC

House sparrow

We have a white sparrow within our estuary, not an albino, just a very pale bird.

One wet sparrow.
Colin M McKenzie

Pheasant

Our pheasant is a lovely old bird which has been here over five years. He is easy to recognise apart from his beautiful colour because he has a limp — he must have injured his leg at some stage. He has a couple of wives and every now and then we see chicks. Unfortunately, being a built-up area with cats and dogs, his territory is getting smaller. The hen birds are reclusive, but our old bird likes to be cock of the roost and will stop, flap his wings and crow outside our window. He and my dog have an understanding: if he is too slow, he will be eaten.

Hen pheasant & chick from our lounge window.
Mik Peryer

Ring-necked pheasant from Mik's lounge window.
Colin M McKenzie

skylark

Twittering on high, the skylarks on our hill are such lovely birds, just hanging there seemingly stationary in the sky. As the area is developed, their habitat disappears.

Peter Morrison, DoC

David Mudge

Swallow

Swallows dive for insects and are superb fliers. Our swallows decided to nest under our eaves. They ducked and dived but couldn't find a decent nook, so I obliged with a flat board nailed under the eaves. They started to build immediately, bringing mud from the river and splashing it everywhere. From my armchair I couldn't see into the nest so I lowered the perch a bit and could sit and watch TV with one eye and the swallows with the other.

However, darker forces were at work. Minty, our cat, spotted the swallows nesting and sat on our television about a metre from the nest. She watched the swallows; they watched Minty. The outcome was an abandoned nest.

A pair of swallows decided to nest in my farmer friend's garage, the doors of which were permanently open. They do tend to make a mess and having his new car pooped upon didn't impress the farmer one little bit. He shut the garage doors, caught the swallows at the window, put them in a box and transported them to another farm by a lake, forty-five kilometres away, where he let them go. It took the swallows just two days to find their way back to the garage and their nest.

Like fantails, swallows will follow cattle into new pasture, catching insects disturbed by the cattle. One pair nested in a farmer's rotary cow bales. As you know, rotary cow bales rotate, so the nest was in a different position as the cows were milked. The swallows always managed to find it.

Dick Veitch, DoC

Peter Blok, DoC

Whistling tree frog

D Garrick, DoC

I found a light brown Australian whistling tree frog in my garage. Apparently they were introduced to Himatangi about fifty years ago, and are now in our estuary. They are small and climb like a monkey, using their front feet like arms to pull themselves upwards. Rather than croak, they whistle.

Stoat

The stoat which was so nosy when the seagull tangled in the fishing line is, with his family, probably having a hard time in the estuary. His main source of food, the rabbit, has had its population decimated with the virus that farmers released into the wild. His food source at the moment is birdlife and mice. When the food chain is interfered with, there are repercussions for the animal and bird population one way or another.

Stoats are strong swimmers and our offshore islands are only predator-free for as long as stoats don't take a fancy to swimming out and eyeing up the birdlife.

Monty the cat

Our cat Monty, a Burmese, is quite a character. One day he came through the cat door with a garden glove in his mouth, which he'd pinched from somewhere — the following day he arrived inside with its mate. The next thing to turn up was an old pair of tracksuit pants. I really don't know how he got them through the cat door. Over a period old towels, T-shirts and rags turned up. It was some time before we found out where he was flogging them — it was from our neighbours three doors down. We came home one day to find Marie, our neighbour, knocking on our door to ask if we had her husband's trousers. We went looking and found them the other side of our fence, snagged on a gorse bush. Monty had gone into their house, up the hall to the bedroom and pinched Marie's husband's best trousers off the bed. He dragged them home to our place. Imagine a cat pulling a pair of trousers through the long grass between our sections. He must have spread his legs and pulled them along. Unfortunately there wasn't a wallet in the trousers or we could have had a handy income!

Cats living in an area like this next to a nature reserve can create havoc with the birdlife, so they must be kept under control and not let out at night. One day Monty went down our hill to the river, found a duck's nest and proudly bought back an egg. This was quite a feat, as ducks have large eggs. He had to stretch his mouth wide to carry it, but he never broke it.

Minty of the swallows, and Monty.

Hedgehog

Hedgehogs live around the lagoons and river on the rough ground and come out at night to feed. Meg will pick them up in her mouth, ignoring the prickles. She brings them to me, pleased with herself, and gets a good telling off, but is reluctant to leave them alone. When approached or frightened the hedgehog rolls itself into a ball with prickles erect. They hibernate in the winter, making a nest of grass around themselves in a nice dry spot under a log or in a depression under a bank. They end up like a football hidden under the foliage until spring.

    This is April who lost her eyes to a magpie and was saved by my friend Edith Crane, who looked after her for seven years, during which time she mated, no matter that she couldn't see, and had twenty-seven babies — averaging five at a time.

    Hedgehogs can range several kilometres at night. If they come to an obstruction — like an ivy-covered fence — and can't find a way around it, they have been known to climb to the top and drop, in a ball, on the other side.

April.
Edith Crane

Edith Crane

Birds noted in the Waikanae Estuary

List compiled 1993–2000 by John Bycroft

Banded dotterel (pohowera)
Bar-tailed godwit (kuaka)
Black-backed gull (karoro)
Black-fronted dotterel
Blackbird
Black shag (kawau pu)
Black swan
Canada goose
Caspian tern (taranui)
Cattle egret
Chaffinch (pahirini)
Dabchick (weweia)
Dunnock (hedge sparrow)
Eastern curlew
Falcon (karearea)
Fantail (piwakawaka)
Gannet (takapu)
Goldfinch
Greenfinch
Grey duck (parera)
Grey teal (tete-moroiti)
Harrier hawk (kahu)
House sparrow (tiu)
Kingfisher (kotare)
Lesser knot (huahou)
Little black shag (kawau tui)
Little egret (matuku moana)
Magpie
Mallard (parera)
Morepork (ruru)
Pacific golden plover
Paradise shelduck (putangitangi)
Pied oystercatcher (torea)
Pied shag (karuhiruhi)
Pied stilt (torea-poaka)
Pukeko (swamp hen)
Red-necked stint
Red-billed gull (akiaki)
Redpoll
Reef heron (matuku tai)
Ring-necked pheasant (peihana)
Royal spoonbill (kotuku-ngutupapa)
Sandpiper
Scaup (papango)
Sharp-tailed sandpiper
Shoveler duck (kuruwhengi)
Skylark
Spotted shag (parekareka)
Spur-winged plover (tuturuatu)
Starling (taringi)
Thrush
Tui
Turnstone
Variable oystercatcher (torea pango)
Welcome swallow
White-eye (tauhou)
White-faced heron (matuku moana)
White-fronted tern (tara)
Wrybill (ngutu parore)
Yellowhammer

UNIDENTIFIED FLYING OBJECT

My friend Eileen Thomas photographed this bird on the wharf at Waimanu Lagoon, March 2004. Can you identify it? If you can please contact us:
mick.moira.peryer@paradise.net.nz

Birds, 2003

Of the birds noted 1969–72 by Wodzicki, these are thriving or have arrived since:

Shoveler duck	~ Two pairs recorded 1962–72. Increased dramatically, breeding on the Waimanu, Waimeha lagoon and river.
Black swan	~ Four recorded 1962–72. They are now almost a pest, with several pairs breeding on both lagoons and the river.
Royal spoonbill	~ Not reported 1962–72. A recent arrival with up to twenty-two birds using the estuary at times.
Welcome swallow	~ Not reported as nesting 1962–72. Now nesting in the estuary — increased dramatically.
Scaup	~ No previous reports 1962–72. Twenty to thirty regularly on Waimanu lagoon with some nesting.
Spur-winged plover	~ Recent arrivals to Waikanae estuary.
Paradise shelduck	~ Sixteen seen 1972 and three pairs on other occasions. Breeding on Waimeha lagoon and seen on the river in large numbers.
Chestnut teal	~ Not reported 1962–72. Four birds sighted 2003.
Brown teal	~ Not reported 1962–72. Now regularly in estuary in small numbers.
Grey teal	~ None recorded 1962–72. Sighted on Waimeha lagoon.
Coot	~ None recorded 1962–72. Two sighted on Waimanu lagoon 2003.
Canada geese	~ None recorded 1962–72. Fifty to over a hundred fly in at times.

Wodzicki KA (1946) The Waikanae Estuary, An ecological survey of New Zealand birds. *Emu 46*: 3-43
Wodzicki, KA et al (1978) Waikanae River estuary, changes to habitat and bird fauna evident from surveys thirty years apart, *NZ J. Zook.* 5: 551-579.

Acknowledgements

How do you write a book of tales like this?

Well, having retired to the beach and being a fellow who is always busy I had this wonderful idea to show this part of the world to tourists. I know my birds — an interest in nature was implanted in me as a youngster by my mother. The books she would buy me were nature books about badgers and moles and stories of the English countryside, as in those days most nature stories were English. I thought that if I was going to show visitors our estuary birds I had better have some stories to tell them — so I penned a little book of my experiences of birds' behaviour.

Having done this, and getting a good response, I thought I would venture forth with a more comprehensive book with photographs of our individual Waikanae birds. When I produced the first little booklet, I had advertising of my bird tours on the rear page — I gave it to my local barber Mike to put in his salon so that customers could read it. He said I would have to give his Rotary Club a talk on the birds and booked me for three weeks' time.

I'm a retired builder and builders are good at driving nails, but I'd never talked to a group before — this was daunting. Mike said 'It's a piece of cake, go to the local shop and buy a slide film for your camera, go to the estuary and snap some birds, get a projector and see you in three weeks.'

Realising this wasn't possible, I put my talk back a few weeks, rang the local library for slides on birds and was put in touch with the Department of Conservation library in Wellington. Their lovely librarian Ferne McKenzie introduced me to many wonderful slides — so I did my talk and showed the slides. It went well. Since then I have got into the swing of things and talked about Waikanae estuary birds to hundreds of people.

To the people who contributed some of these stories, I am forever grateful: Brian Atkinson, Collis Blake, Tom Caithness, Kay & Trevor Campbell, Betty & Don Logan, Erwin & Carl Lutz, Jamie Ritchie, and Shelia Rumsby.

To the people who contributed photos: Colin McKenzie, Eileen Thomas, David Mudge, Alex Scott and the Department of Conservation, my heartfelt thanks.

To Norma McCallum and Wendy Eriksen go my thanks for their input into the little booklet which has been incorporated into this book. Thanks also to Grace Suckling and Ralph Powlesland for their knowledge of the nature of things.

I would like to acknowledge Chris Budgen who, over 30 years ago as an English immigrant, was on a family outing to Greytown and stopped by the river and started taking photographs. I had lived in this area over many years and couldn't see what he was photographing.

"Look at the beauty of those mountains!" he said.

I knew the area intimately, yet had always been too busy as a youngster to see the beauty we in New Zealand take for granted… From that day to this I stop every now and then to take in the wonderful scenery that is New Zealand. Thanks, Chris.

To Charles Lloyd, my long-suffering neighbour, his wife Eth and son Nick, I owe a great deal for their computer skills, humour and patience.

Last and most important, I thank my darling wife Moira. She has had to put up with much as I have worked through to the conclusion of this book.

Waikanae Beach on a busy day.
Eileen Thomas

Tour beautiful Waikanae estuary

- 2-3 hour birdwatching tour with your personal guide.
  Stroll along the sandspit/riverbank enjoying magnificent scenery and the variety of waterfowl which inhabit this wonderful area. (Approximately 60 species of birds visit the estuary at times)
- Visit Mirek Smisek and Pamela Annsouth's famous pottery. Then morning or afternoon tea at home with your guide.

> Kapiti Bird Tours
> 160 Weggery Drive West, Waikanae
> Phone (04) 905 1001
> Mobile 021 750 603
> mick.moira.peryer@paradise.net.nz
> www.kapitibirdtours.co.nz